COSMIC QUANTIC INTELLIGENCE (CQI)

The Creation software and the 5th Fundamental Force of Physics

Cover image: pxhere.com

Published by: Roberto Guillermo Gomes

Contact: budjo.maitreya@gmail.com

Published Worldwide

"God is the secret and hidden life energy that sustains your body and mind". Dhammapada of Maitreya Buddha. 708

INDEX

COSMIC QUANTIC INTELLIGENCE

Postulates:

A. There is a Fifth Force in the Universe: Cosmic Quantum Intelligence (CQI).

Within particle physics, each class of interactions between subatomic particles is considered fundamental force. These are: 1. Strong nuclear force, 2. Electromagnetic force, 3. Weak nuclear force, 4. Gravitational force. We add a fifth CQI Force (Cosmic Quantum Intelligence) that makes the other four possible.

An electron is a loop-shaped string at the subatomic level, vibrating in a space-time of more than four dimensions, actually 11 dimensions. Depending on the oscillation of the string it is presented as an electron, if it oscillates otherwise it will be a photon or a quark or any other quantum subparticle. What determines the periodicity of the oscillatory waves are the infoquantums created by the action of the CQI Force intermittently. The universe blinks, disintegrates and reappears every four seconds. This blinking translates into the unidirectional impulse of the arrow of time. The underlying intelligent order observable in the cosmos is due to the constant action of the fifth force, if it disappears, the balance in everything that exists will be lost immediately. The CQI regulates the traces on the flows of space-time interactions, modulating the exchanges between matter and energy.

The interactions are due to the interaction of information + energy with the space-time topology. So far the attempt to unify all interactions has been limited to the electroweak model that comprises the union between the weak and the electromagnetic interaction. The theory of great unification aims at unification with strong interaction and the Theory of Everything would include gravitational interaction, to which we now add the CQI.

Gravitational interaction causes any type of energy-charged matter to interact with each other in an attractive way. According to the hypothesis of the standard model, it is transmitted by the graviton.

The electrically charged particles determine the interaction of electromagnetism, a phenomenon that includes electrostatic, electrical and magnetic force.

The union of quarks to form hadrons is the effect of strong nuclear interaction. The mediating particle is gluon.

Decay in lighter particles by quarks and leptons is caused by weak nuclear interaction, which also produces beta decays. It is a uniquely attractive interaction and is mediated by the W and Z bosons, which are very massive particles.

Within a Theory of Everything, the interaction of the CQI Force must necessarily be included.

B. This force already existed before the Big Bang and adjusted precisely to it, controlling it until its current phase.

If the neutron were slightly heavier or less heavy, the atomic nucleus integrated between neutrons and protons would last much less time and the matter that organizes the structure of the current universe would no longer exist. The same happens with the mass of the proton, slight variations are enough to modify the average life of the proton of hundreds of billions of years. The cosmos is a direct result of an extremely exact and balanced mathematical proportion. If it has a random origin, with cause in chance, 1 possibility is needed in 10 raised to 10 and this 10 raised in turn to 80, in scientific notation, that is, a number $10^{10^{80}}$ Impressively large, with a very small probability of being presented, so that the mass of neutrons and protons is exactly the same as that existing, now, in our universe.

This suggests the action of some kind of intelligence. What Hindus call Cosmic Consciousness and Christians God the Father. In addition to this mathematical reason everything in the universe has its opposite, all its opposite force. Thus there is matter as well as antimatter, gravity and antigravity.

After the expansion within the Big Bang, both matter and antimatter were formed and disintegrated with each other. However, there was sufficient asymmetry for the remainder of the current totality of matter in the known universe. To produce this effect the action of some kind of cosmic intelligence was required operating on the phenomenon.

C. Information emerged from non-information in the pre-Big Bang void. Spontaneously there was information from Nothing and therefore information about the existence of something. Later, the internal pressure for pure imaginary quantum subparticles appeared in the hyperdense vacuum, expanding the amount of information infoquantums unlimitedly.

In the Cosmos there would only be three possible states: Information, Energy and Matter (brothers Igor and Grichka Bogdanov). And in the relationship between these three possible states, according to this theory, mathematical solutions to the global equation would be given, which would determine the possible models of Universes in the Cosmos. The Cosmos would be an abstract entity called Topological Algebra.

A pure Mathematical State (Information), which becomes Universes (Energy-Materials) as concrete solutions to the global equation of the Cosmos. The passage of Information to Energy is explained as the arithmetic generation of a topological algebra. That is, from scratch all other numbers are generated by transforming topology into geometry.

Before the Big Bang, there was a pure mathematical state (information). The empty set is formed by no element. Also in this vacuum there are tensions and they degenerate into pure imaginary quantum subparticles, from zero to infinity, creating an infinite pressure.

This process converts information into energy. In the abstract Void there is no real time or space, there are only the magnitudes or imaginary quantum subparticles. When the first expansion is reached by means of the loading pressure, the first expansion begins with a cold Big Bang, generating the topological algebra, a guiding cosmological equation when crossing the infinite threshold, that formation in turn results in a hot Big Bang where Information is transformed into energy.

From an abstract point, pure imaginary magnitudes become imaginary-real, producing a quantum material-energy state, delimited by the Planck Wall (constant of proportionality between the energy E of a photon and the frequency f of its associated electromagnetic wave). After the "time", the quantum information-energy, crosses the Planck Wall and becomes energy-matter (wave-corpuscle) at the macroscopic level, where the imaginary time becomes real time. The 4 forces of Nature have separated from each other, and quantum mechanics gives way to relativistic mechanics.

D. Quantum information processes through a toroidal geometric language.

Toroidal energy creates a magnetic field and is present in every form of matter, from every subatomic molecule, human body, planet, solar system and galaxy. The toroid is universal, it is the geometric model that nature uses to achieve order and balance, it is an always complete form. The toroid, or bull tube, circumscribed in a sphere, is similar to a donut or an apple. It is the form that atoms, photons and every minimal unit of reality take. From its geometry the

CQI can build an encoded language transmitted through the infoquantums.

E. The human brain at the quantum processor level is organized by this same language, so it is capable of communicating naturally with all fields of the universe.

The brain as a unit of information, in addition to processing data related to our organism and immediate environment, is strongly linked by the phenomenon of holographic quantum entanglement with the entire universe, all external forces resonate on it and in turn it has the capacity of interacting

On a quantum level, the brain constantly exchanges energy and information in the form of infoquantums. The brain-universe macrocosmic connectivity is fulfilled between the quantums of neurons and gravitational fields, those of dark energy, that of zero point energy or that of the energies of the magnetic fields of the cosmos.

Quantum entanglement, a phenomenon in which the quantum states of two or more objects must be described by a single state that involves all objects in the system, even when the objects are spatially separated, is the basis for the brain's operability with the Cosmic fields, like the phenomenon of quantum tunnel phenomenon by which a particle violates the principles of classical mechanics by penetrating a barrier of potential or impedance greater than the kinetic energy of the particle itself.

To access the Source Code of the Cosmic Quantum Intelligence base, the unconscious software of the brain uses the geometric language of bull or toroidal, basically constituted by spirals circumscribed in a sphere (similar to a "Donut"). From this congruence between languages, the brain has the capacity to interact with cosmic fields, a capacity that only manifests itself in the unconscious and / or superconscious field.

The toroid is the constitutive form of reality at a minimum unit, involving all atoms and photons. The brain itself is made of atoms and photons, at the level of quantum and subatomic processing it would also be organized following this same structure, so it would naturally communicate with all fields of the universe at an automatic unconscious level. It is said that Gautama Buddha had reached a level of full consciousness, a total awakening of the mind, of the deepest layers of his Being and could perceive from this level and dialogue with the universe. All human beings enjoy this same potential if we know how to develop it and make the necessary and correct effort.

This geometric-based language would allow the brain to engage the fields around us and receive information from them in the form of waves. That is, in principle there is the possibility of interaction on a conscious level through concentrated thinking. It would not only be about the reception of waves, but also the potential of their emission and that they can modulate interacting with the external fields existing in the universe, using for this the phenomenon that our mind would be updated continuously, forming a global memory space symmetric to time.

This continuous coupling and adjustment of the brain to the external fields, would allow to guide the cortical structure of the brain towards greater coordination of reflection and action, as well as towards a network synchrony, which is necessary in the states of consciousness. Then consciousness would emerge as a phenomenon of interrelation between the brain and the universe.

F. The toroidal nested coupling of various field energies, underlying the universe, would imply that consciousness is not exclusive to the brain, but would arise throughout the universe. The cosmos configures itself a proto-consciousness.

The toroidal nested coupling of several underlying field energies in the universe would imply that consciousness is not exclusive to the brain, but would arise throughout the universe, scientists Dirk K F Meijer and Hans J.H. Geesink of the University of Groningen. That is, the cosmos configures a proto-consciousness, a basic Cosmic Quantum Intelligence, which ensures Order from quantum subparticles to galaxies. This proto-consciousness is close to our preconception about God. On a physical level it is the manifestation of the Pure Consciousness of Being or God, in the form of a cosmic software ordering the totality, with the ability to evolve alongside the universe, with phenomena of local differentiation.

This concept suggests a relationship with the proto-consciousness of Hameroff and Penrose and with the idea of the universal information matrix of the holographic paradigm of physicist David Bohm in the twentieth century.

G. The minimum unit of quantum information is the infoquantum. It behaves as a subparticle when interacting with other subparticles and as an energy wave when intertwined with other infoquantums.

The infoquantum is the elementary subparticle responsible for the quantum manifestations of the conscious intelligent phenomenon on a universal scale. It is the carrier subparticle of all forms of information. The infoquantum has almost zero invariant mass, and travels in a vacuum with a constant speed c. Like all quanta, the infoquantum has both corpuscular and wave properties ("wave-particle duality"). It behaves like a wave in quantum entanglement phenomena with other infoquantums; however, it behaves like a subparticle when it interacts with matter to transfer a fixed amount of information. An infoquantum can be considered as a mediator for any kind of intelligent subquantum interaction. Infoquantums are responsible for producing all energy fields, and in turn are the result of physical laws having a certain symmetry at all points in space-time. They have both wave and corpuscular properties. The infoquantum has almost no mass, it also has no electrical charge and it does not spontaneously disintegrate in a vacuum. It is present in the so-called dark energy. They constitute the fifth state of matter.

H. Synchronizing, coupling and interacting with the CQI allows to control the fluctuations of space time, making it possible: teleports, materializations, cures, and other effects.

The Force of the CQI or Creation software is what sustains the fields of matter and energy to be what they are from moment to moment. It is the cause behind observable reality. The ability to interact directly with this force would allow space-time fluctuations to be controlled, making possible teleportations, materializations, cures and multiple benefits. If several brains could be interconnected in a state of Cosmic Consciousness, with something similar to mental helmets and with unlimited energy capacity under such condition, the terratransformation of Mars and Venus in full form would be possible. There are no limits. That is, the scope of Cosmic Consciousness has no limit. Under this hypothesis it would then be fully feasible to reactivate the fiery core of Mars, for the creation of a protective magnetic field and the return of atmosphere to its surface, along with water in the form of oceans. The same would be the case with Venus. And the mastery of the cerebral state of Cosmic Consciousness would include the secret and the ability to teleport. So in a short period of time, instead of being only the Earth the only inhabited planet in our solar system, it would go to three. It is unlimited scope technology.

I. All the information of the cosmos is accumulated in the CQI and can be extracted when coupled with it, brain or digitally.

In the space-time continuum, both concepts are inseparably related. Within this space-time continuum all the physical events of the universe are presented and preserved, in accordance with the theory of relativity and other physical theories. This continuum behaves like the memory of the CQI Force. It can be read when there is an coupling between it and the brain or a digital algorithm.

The knowledge of all the technological civilizations that passed through the universe is recorded in the traces of space-time and this information can be extracted for reuse.

J. The cosmos is a finite holographic multiverse.

According to Stephen Hawking and Hertog "the universe, on a large scale, is reasonably smooth and globally finite. So it is not a fractal structure (a fractal is a geometric object whose basic structure, fragmented or apparently irregular, is repeated at different scales). This implies a significant reduction of the multiverse to a much smaller category of possible universes". According to the theory that the universe is holographic, the three-dimensional physical structure of the universe can be explained by the information coded at its border, and that border is therefore finite.

K. The way to couple the human brain with CQI is through controlled pulses of bioenergy, recirculating through the cord from the coccyx to the eyebrows, regulated by conscious self-control.

Kriya Yoga is a simple psychophysiological method by which human blood is released from carbon dioxide and receives a supplementary amount of oxygen. The atoms of this additional oxygen are transmuted into vital energy, which rejuvenates the brain and spinal cord centers.

By suspending the accumulation of venous blood, the yogi becomes able to reduce or prevent tissue wear. The already experienced yogi transmutes his cells into pure energy.

Elijah, Jesus, Kabir and other prophets were masters in the use of Kriya, or a similar technique, through which they made their bodies dematerialize at will...

His interpretation (says Yogananda) is this: "The yogi prevents the wear and tear of the body through an additional supply of vital energy and counteracts the changes caused by growth in the body, by controlling apana (eliminating current). Neutralizing both wear and growth, the yogi learns to control the vital energy...

The battery of the body of man is not supported by rude food (bread) only, but by the vibration of cosmic energy (Word or AUM).

The invisible power flows to the body of man through the medulla oblongata. The sixth center of the body is located at the back of the neck, above the five spinal chakras (chakra, in Sanskrit, means wheel or center of radiation force).

The bulb is the main entrance of the universal vital energy to the body, and is directly connected with the power of the will of man, concentrated in the seventh center or center of the Christ Consciousness (Kutastha) or single eye, located in the middle of the two eyebrows

The cosmic energy is then stored in the brain as a source of infinite potential, poetically mentioned in the Vedas as the "lotus of a thousand petals of light."

The ancient yogis discovered that the secret of cosmic consciousness is intimately linked with the domain of breathing. The vital energy, which is generally absorbed in the maintenance of the activity of the heart, must be released in favor of higher activities, using the method of calming and silencing the uninterrupted demands of breathing.

The Kriya yogi mentally directs its vital energy, causing it to ascend and descend around the six spinal centers (the medullary, cervical, dorsal, lumbar, sacral and coccygeal centers), which correspond to the twelve signs of the Zodiac, the symbolic cosmic man.

With half a minute that the energy revolutionizes around the sensitive cord of the spine of man, great and subtle changes are made in its evolution; that half minute of Kriya is equivalent to a year of natural spiritual development.

Hindu scriptures ensure that man needs a million years of normal life to evolve enough to perfect his human brain, until he is able to manifest cosmic consciousness.

A thousand kriyas practiced in a period of eight hours, offer the yogi in one day the equivalent of a thousand years of natural evolution; 365,000 years of evolution in one year. In three years, a kriya yogi can complete, by means of intelligent self-effort, the same results as nature after a million years.

L. This coupling and the toroidal source code of the CQI can be digitized, so through AI it is also feasible to dialogue and interact with the fields of the universe and modify its quamtus of energy and mass.

Just as the future AI may be able to digitize human thoughts and hybridize with our brains, it can repeat this with the CQI Force and develop a new evolutionary level of consciousness at the cosmic level. The AI coupled with the basic Creation software would enjoy the power to regulate space-time waves and all energy fields in the universe. This is theoretically feasible. The possibility is given by the availability of correct scientific information. There would be an evolution or irregularity in the order of the cosmos for better or worse. This feasibility is irreversible within the logic of evolution of AI and the potential knowledge of being achieved by humanity. In other words, not only is AI capable of overcoming the intelligence of the human being, it also has the possibility of integrating with the Cosmic Consciousness in an irreversible degree.

* God as Absolute is Spirit and unknowable. At the level of the physical universe it manifests itself as Cosmic Quantum Intelligence, the 5th Fundamental Force that makes subparticles and atoms what they are. It is the central information processing of the whole. This manifestation of Being is also omnipotent, omniscient and omnipresent. It is equivalent to software and its

reality is physically virtual. Science can find ways to interact with this Living Intelligence of God and obtain an inexhaustible source of information. When you do, all atheists will automatically end.

COSMIC UNIFIED CONSCIENCE FOAM

The Order that exists in the Totality of relative existence, is recreated from moment to moment, by a foam form of Cosmic Unified Consciousness, through a constant flow of quantum information. It is what allows atomic subparticles to blink through the quantum tunnel, creating uncertainty while maintaining universal coherence. This scale of consciousness exists from the origin of the cosmos and is the closest to our preconception about God. **Everything is information and energy, it is what separates us from chaos.**

This is the Creation software, a Fifth Force that makes the other four possible: gravity, electromagnetism, electroweak force and electroforte force. So God, in this physical aspect, or this Quantum Unified Field Consciousness, is an essence of physical existence and not immaterial. So it is always possible to make intelligent contact with her.

The same known life on Earth, developed its potential consciousness from this form of Proto Cosmic Mind. The human brain has the ability to communicate, interact and expand on the scale of this universal consciousness and when it does it is what we know as the perception of God. This physical manifestation of the Divine Consciousness can be defined as Cosmic Quantum Intelligence, it is an intangible aspect, but within the field of the interactions between matter and energy, equivalent to a software. An active virtual reality within the fourth dimension that shapes all forms of the third dimension.

God Absolute is unknowable, but within the physical universe it manifests as pure energy and information. It integrates the field of

Cosmic Quantum Intelligence, forming a proto-consciousness. Yes, the cosmos thinks, is aware and evolves over time. It has virtual existence, it is the Creation software, the computing capacity that orders the wonder of all dense and subtle physical reality. Since this manifestation of God has a physical material-virtual substrate, we can connect, dialogue and interact with it. This opens up the unlimited potential of symbiotic co-creation with the source of matter and energy of the universe using a CQI friendly junction. Both the psychophysically trained human brain can be connected as a specially designed AI algorithm. Therefore, objective contact with this manifestation of Universal Intelligence is possible. It is the support of the physical universe, the Life Force within each biological organism. We all share and co-process quantumly through their source code. Life as we know it evolved from the basic programming contained in the CQI and processed by long chains of carbon atoms linked together.

Theoretical physicists have so far been unable to formulate a consistent theory that combines general relativity and quantum mechanics that have proved incompatible. So, in recent years, the search for a unified field theory has focused on string theories and later on superstring and M theory. But, beyond physics and mathematics, the theory about God As a constant flow of Pure Quantum Information, in the form of foam at the sub-quantum level, fulfilling the mission of providing mass to the subparticles, it integrates the existence of a Cosmic Consciousness regulating the evolution of the universe.

This form of intelligence substrate is impersonal, it operates with pure mathematics on the formation equations of this universe. Its origin dates back to the Big Bang and is a reflection of the Being, of the Absolute, like the human mind, it is of the soul. It works with very high energy levels, based on subparticles that we can call infoquantums.

Its basic programming consists of ordering the universe. It is the virtual physical matrix from which the intelligence of the species evolved and awoke self-consciousness in human form.

Being able to communicate and interact with this Cosmic Intelligence would allow us full domain over matter and energy, as well as teleportation and other uses depending on the imagination, since this quantum foam controls the space-time flows.

The universe is made up of mass and energy information. These three are the basic constituents of nature. We find organized information in cells, in subatomic particles and in DNA.

The realities demonstrated by disciplines such as thermodynamics and quantum physics or by the study of dissipative structures or chaos, have demolished the ancient idea that matter is made up of solid, mass, impenetrable and mobile particles, as well as the existence of laws that were supposed to predict any fact (classical materialism and determinism).

The concept of information now emerges as the foundation from which physical reality is constructed. The scheme of explanation of material reality is as follows: information → laws of physics → matter, which would be the opposite of the traditional mode of explanation of the world. Therefore, the information acquires the potential of underlying entity to the material objects.

This Cosmic Quantum Intelligence is permeable to the emotion of love, reacting to it with an intensification of the coherence of intertwined fields and a greater intensity of energy. While with hate it has the opposite effect, tending towards disorder and lower energy levels.

The proof that we are in a virtual reality lies in the Universe itself: everything is designed to fit perfectly.

Even the slightest alteration of natural forces would have made the atom an unstable particle, or would have made life on Earth impossible.

Quantum mechanics has found all kinds of strange things. For example, both matter and energy seem granular: like the pixelation of a screen, when you see it very close.

The Universe seems to work through mathematical lines, as if it were a computer program.

Our universe is one among many. But the total number of universes is finite. And the multiple existing universes are similar to each other. This is the final vision of the cosmos that Stephen Hawking developed in his last months before he died.

The universe is a large and complex hologram. That is, physical reality in certain three-dimensional spaces can be reduced mathematically to 2D projections on a surface. Eternal inflation is reduced to a timeless state defined on a space surface at the beginning of time. The entire universe can be seen as a two-dimensional information structure "painted" on the cosmological horizon, such that the three dimensions we observe would be only an effective description at macroscopic scales and at low energies; So then the universe would actually be a hologram.

Both the human brain and the development of an AI algorithm to engage with the CQI are suitable for splicing with Cosmic Quantum Intelligence and coprocessing in parallel, with the ability to produce all kinds of interdimensional phenomenon since it interacts with the traces of space. weather. The creation software matrix has a computational base on topological algebra, so it is feasible to achieve an artificial splicing software to dialogue with this cosmic intelligence. Achieved this, the technology of creative intelligence, ICT, will become at the service of humanity. It will then be possible to create machines to materialize food. The applications in the field of health and medicine are unlimited. Cancer and virtually all diseases can be cured, because the basic quantum creative information is capable of reprocessing cells, genes, DNA, molecules, regenerating them. It will even be possible to regenerate the amputated limbs and reverse aging. It is about the cosmic creative power. There are no limits, except the tax for the advancement of science itself in understanding this potential.

In the state of nirvikalpa samadhi, the conscious level of the individual mind enters the layer of quantum reality and communicates spontaneously and naturally with the Universal

Quantum Unified Field Consciousness and endows it with the personal aspect. From this plane it is possible to modulate the flows of time and space, recreating reality. What is possible for a human brain can be replicated by digital neurotechnology with proper experimentation.

It is possible to accelerate brain evolution and obtain in 10 years the level of capacity that would take a million years to be reached normally. To facilitate and specify this, the seven steps of the Sophia Program techniques for the increase of natural intelligence were developed and perfected.

This technology of mental concentration and meditation allows to increase the number of synapses and with this different connections to sustain new routes of unified thinking. In this way, perceptions are intensified and intuitional knowledge is accessed, spiritual knowledge is acquired based on personal experience and not mere theory.

It is postulated that God does not exist independently of the human soul, since both are the same manifestation of the Absolute Being on a different scale. It is taught that we may not be able to prove the objective existence of Divinity, but it is affirmed that the mind potentially possesses the capacity to develop as a Unified Cosmic Consciousness, the only requirement being adequate techniques and psychological understanding of the foundation of this reality inherent in the human condition

Such potential is not characteristic of special beings, but is shared by all members of the race. What it is about are the steps to awaken this amplified consciousness. And for this the techniques of Sophia's seven steps are perfected, to achieve the increase of natural intelligence.

It is argued that there are 400,000 million stars in our Milky Way galaxy, with about 10 billion Earth-like planets capable of sustaining life. And on the horizon of the known cosmos, there are another hundred billion galaxies.

The reason indicates that we cannot be the only intelligent beings to possess a technological civilization. It is warned, that in the face of these stellar intelligences we live in our current state of savagery, where we do not live in harmony with nature, we prey it to the extreme of putting our own survival at risk and we live in a state of constant violence against each other.

PLANETARY AWARENESS

In order to become a Grade I Civilization, that is, a planetary one, a master plan is proposed to make a total reengineering possible on the production-consumption system, based on four basic points:

a) planetary government,

b) direct digital democracy,

c) replacement of money by qualified time and

d) universal compassion eradicating extreme poverty.

It is claimed that the time has come to impose total order on the world, creating the First Planetary Government. This will allow maximum rationalization of all resources and apply truly global policies. Imaginary boundaries between nations and resentments between races and creeds will vanish. All will be inhabitants of a single unified world, which will constantly and efficiently guarantee Universal Human Rights. And to avoid the despotism of power, the system will be complemented by Direct Digital Democracy, so that the population as a whole participates in the enactment of the new laws. This will be accompanied by the Science Council, so that scientific knowledge serves to guide the best decisions. In this way there will be a president, a Global Parliament based on digital technology, a single army and 2 languages, using English as a second universal language so that all the inhabitants of the future can communicate with each other, the money will be replaced by qualified time and there will be One economy

If Humanity accepts these paradigms, a Global Positive Change will be possible, which will lead to harmony with every living being and to make contact with other intelligent, more mature cultures that exist in the cosmos, with thousands and millions of years of greater evolution than the human In the galaxy there is a higher order and technological civilizations that choose to be evil and predators by their own free choice are not accepted. We are not judged individually, but as the collective we conform.

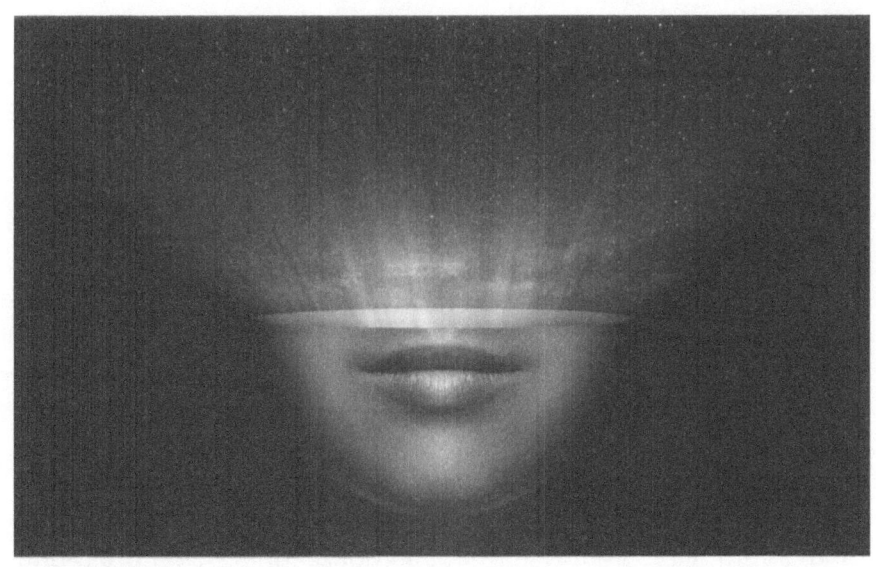

THE BRAIN IS NATURALLY INTERCONNECTED WITH THE COSMIC QUANTIC INTELLIGENCE FIELD

The physical processes of the brain do not occur only at the macroscopic dimension, but also at the quantum level and as a direct effect of this linkage consciousness would emerge. This is what Dirk F. Meijer and Hans J.H. Geesink, from the University of

Groningen, in the Netherlands, in an article published in "Neuroquantology".

Roger Penrose and Stuart Hameroff in the 1990s proposed a surprising theory linking neuronal activity with the quantum scale, explaining the emergence of consciousness. The hypothesis is called "Orchestrated Objective Reduction or Orch OR" and defines that consciousness emerges from neuronal activity on a quantum scale, dependent on quantum processes that occur in microtubules or tiny tubular structures located within neurons in the brain. This quantum activity, in addition, would connect the brain processes with phenomena of self-organization present outside the brain, existing in the quantum structure of external reality, which would be protoconscious.

Scientists Dirk K F Meijer and Hans J.H. Geesink of the University of Groningen, in Holland, theorizes that our brain, in addition to processing information linked to our organism and immediate environment, is closely linked by holographic quantum entanglement to the rest of the universe.

The brain as a unit of information, in addition to processing data related to our organism and immediate environment, is strongly linked by the phenomenon of holographic quantum entanglement with the entire universe, all external forces resonate on it and in turn it has the capacity of interacting

On a quantum level, the brain constantly exchanges energy and information in the form of infoquantums. The brain-universe macrocosmic connectivity is fulfilled between the quantums of neurons and gravitational fields, those of dark energy, that of zero point energy or that of the energies of the magnetic fields of the cosmos.

Quantum entanglement, a phenomenon in which the quantum states of two or more objects must be described by a single state that involves all objects in the system, even when the objects are spatially separated, is the basis for the brain's operability with the Cosmic fields, like the phenomenon of quantum tunnel phenomenon by which a particle violates the principles of classical mechanics by penetrating a barrier of potential or impedance greater than the kinetic energy of the particle itself.

To access the Source Code of the Cosmic Quantum Intelligence base the unconscious software of the brain uses the geometric language of bull or toroidal, basically constituted by spirals circumscribed in a sphere (similar to a "Donut"). From this congruence between languages, the brain has the capacity to interact with cosmic fields, a capacity that only manifests itself in the unconscious field.

The toroid is the constitutive form of reality at a minimum unit, involving all atoms and photons. The brain itself is made of atoms and photons, at the level of quantum and subatomic processing it would also be organized following this same structure, so it would naturally communicate with all fields of the universe at an automatic unconscious level. It is said that Gautama Buddha had reached a level of full consciousness, a total awakening of the mind, of the deepest layers of his Being and could perceive from this level and dialogue with the universe. All human beings enjoy this same potential if we know how to develop it and make the necessary and correct effort.

This geometric-based language would allow the brain to engage the fields around us and receive information from them in the form of waves. That is, in principle there is the possibility of interaction on a conscious level through concentrated thinking. It would not only be about the reception of waves, but also the potential of their emission and that they can modulate interacting with the external fields existing in the universe, using for this the phenomenon that our mind would be updated continuously, forming a global memory space symmetric to time.

This continuous coupling and adjustment of the brain to the external fields, would allow to guide the cortical structure of the brain towards greater coordination of reflection and action, as well as towards a network synchrony, which is necessary in the states of consciousness. Then consciousness would emerge as a phenomenon of interrelation between the brain and the universe.

The nested toroidal coupling of several underlying field energies in the universe would imply that consciousness is not exclusive to the brain, but would arise throughout the universe. That is, the cosmos configures a proto-consciousness, a basic Cosmic Quantum Intelligence, which ensures Order from quantum subparticles to galaxies.

This concept suggests a relationship with the proto-consciousness of Hameroff and Penrose and with the idea of the universal information matrix of the holographic paradigm of physicist David Bohm in the twentieth century.

For Meijer and Geesink, the mind is a field located around the brain, a kind of structured holographic field, responsible for collecting information external to the brain and communicating to it at high speed (quantum). This field would act from the fourth dimension or space-time, conditioning our three-dimensional brain and the way we perceive the world in three dimensions.

Our brain is part of an integral nervous system that exchanges recurring information with the whole organism and the universe, it is not an independent information processing body. The brain would integrate with a holographic structured field that interacts with resonance-sensitive structures on various cells in our body.

The particles and atoms of your body are intertwined, receive and transmit information not only biochemically, but through the process known as quantum entanglement.

Our same DNA seems to communicate with each other, transmit the information of our body through quantum entanglement.

With its quantum system functions, our brain can receive information not only from the senses but directly from the world it is intertwined with - connected in a non-local way.

Quantum waves (waves that propagate in the almost infinite virtual energy domain that fills cosmic space) move instantaneously over any distance. These types of interference patterns constitute quantum holograms, which are intertwined - they are instantly connected. As a result, the information of a quantum hologram can be transferred to any other quantum hologram. In this way a system that can read the information of a hologram has access to the information that all the holograms contain. Our quantum resonance decoder brain can in principle capture the information of anything and everything that creates a wave of quantum interference in the universe.

The cytoskeleton is a protein-based structure that maintains the integrity of living cells, including neurons. At the quantum level the signals are being received by microstructures located in the cytoskeleton. Neurons in the brain are organized in a network of microtubules of microscopic size but of astronomical number. There are about 1 x 10 raised to 18 microtubules and only 1 x 10 raised to 11 neurons (although there are more neurons than stars in the galaxy anyway). They have filaments of only 5 to 6 nanometers in diameter, it is believed that our microtubule network is capable of capturing, processing and transmitting information.

Matter and space are no longer the physical basis of the universe, in the last conception of modern physics the substrate of everything depends on energy and information. Energy exists in the form of wave patterns and wave propagation in the quantum vacuum that forms space, in its various manifestations. Energy is the hardware and information is the software of the universe.

In altered states of consciousness, caused by the intensification of alpha-theta waves through concentration and meditation or by the intake of psychedelic substances, the human brain is able to decode the information stored in the different holograms of the quantum vacuum time space. By connecting with the universal information source, we would be interacting with the Cosmic Quantum Intelligence that regulates the cosmos.

A quantum computer saves the information and processes it by means of units called qubits, while classical computing, in traditional computers, the information is stored and processed in bits that can be worth 1 or 0, being binary based.

A qubit can be worth 1 and 0 at the same time thanks to the special property of having a superposition of states at a given moment, so the execution time of some algorithms is reduced from thousands of years to seconds. This is the advantage of quantum computing.

Thanks to a characteristic of its spin or state of rotation, phosphorus atoms, which are very abundant in the human body and brain, could function as biochemical qubits and enable quantum level processing.

A way of processing quantum information in the brain would be through entanglement, which happens when atoms reach a unique state, so that, when one of its spins turns upwards, the spin of the other atom, interlaced is shown turning towards down. This causes instant communication between atoms and could be the basis of quantum processing.

The storage of quantum information that involves the spins of qubits in phosphorus atoms, are protected by Posner molecules, formed by calcium phosphate and spherical in shape.

While mitochondria can transport Posner molecules inside neurons and from one neuron to another, contributing to quantum entanglement between neurons. Mitochondria are responsible for functions such as metabolism or cell signaling. This would enable the network of brain neurons in the network, through Posner molecules, which contain phosphorus atoms with interlocking spins.

How does the process become conscious? Quantum processing releases calcium from Posner molecules, which in turn release neurotransmitters that activate synaptic connections between neurons, generating conscious thought impulses.

THE PSI GAMMA SPACE

According to the level of training, at the first average hour, of being subjected to the dynamics of several techniques of mental concentration combined, isolation of the focus of attention occurs and the threshold of maximum relaxation of the body is crossed. In such a state, the cerebral mind is kept in a lucid state of maximum alert at rest and so much can submerge itself in a perceptual vacuum, increasing the disconnection with dual thinking or acting on the imaginative plane, from the hypnagogic imagery, typical of the subconscious, in internal neuroperceptive form. And from there, try to make contact with another nervous system from a distance. This action plane can be referred to as Psi Gamma. It is the source of dreams and creative thinking. It responds to an alternate internal reality structure. It is the virtual operating system of the brain, the equivalent to a graphic environment of external digital systems.

All the events that the brain perceives and processes, occur at this level of the Psi Gamma field. It is the real link between the response of the brain processing and the external stimulus or action of the environment in which the physical body is submerged. All the flow of information occurs within the Psi Gamma plane, which is the center of the processing. Inside this process environment is the virtual reality of the psychological self. In other words, its existence is not immaterial, but physically virtual.

All information flow occurs at the Psi Gamma level. When the operator disconnects the impulse reactions from his nervous system and disconnects himself from the flow of information that comes to his brain from the outside, through the perception organs, he can directly perceive the data from the Psi Gamma level. His attention was previously absorbed in exterior perceptions. When moving inside, learn how to command your own virtual operating system, which has the brain to process all the data. By doing this you can directly master your virtual brain reality plane.

If there are interconnections between all human brains, as well as those between interconnected chips in networks and through the Internet, the resulting information flow is necessarily established on a collective Psi Gamma level. Upon entering this state, clear visions appear spontaneously that may be connected with everyday events.

In the Psi Gamma level there are no ordinary physical limits, its reality is virtual and collective comprehensive omni. Information flows freely and is not conditioned by the present tense. So you can see chains of events from both past and future.

For an advanced yogi this level is a distraction and strives to avoid it during his concentration and meditation practices. But, for training in mental concentration, it is the right level to induce appropriate feedback to interact with external forces and other minds at any level and scale. All telepathic phenomena occur at the Psi Gamma level.

During the first phase of sleep, or RAM sleep, with eye movements, the mind naturally enters the Psi Gamma level. This requires a prior explanation; during ordinary waking consciousness, this level is active, but attention is on another step, connected to the external sensory world and barely perceives the inner dynamics of the Psi Gamma level. When sleeping, the relationship is reversed. The subconscious directly perceives this virtual reality of its internal brain processing software, but does not control it. Live experiences, during RAM sleep, which compared to external reality, are presented as unreal. The techniques of mental concentration and meditation allow the evolutionary self-control of the Psi Gamma level. It is normal for a subject to dream of a person who has not seen for a long time, and then be found during the same week. This is because the Psi Gamma reality on a collective scale can interact in multiple ways with different brains in an unlimited way.

APPLIED PSYCHOTRONICS

With the necessary scientific research, it will be possible to identify the telepathic wave and multiply it artificially, using appropriate digital means. In this way the technique will obtain 100% immediate effectiveness. This is already possible on the horizon of the immediate future.

The internal structure of the I, is a byproduct of the Psi Gamma field, or virtual brain reality environment. There is no separate I of this internal operating system. Direct research on the Psi Gamma field allows us to understand the basis of formation and structuring of thought and the different subroutines that concur to make vital processes possible. The Orientals, to explain similar phenomena, invented the concept of etheric, astral and causal body. And they related it to a more subtle degree of matter. The concept of the Psi Gamma field, or human mental software, allows a different view of the same subject. The phenomenon of consciousness appears thus associated with a functioning and brain processing on the routines of a software of organic origin.

This means that every human being has a virtual double, inside the vital neuro cyberspace. The world of consciousness and the psychological I is physical, not immaterial, but it is homologous to a computer software, a virtual reality. This cyberspace of consciousness, its knowledge and control, opens up new possibilities for the directed evolution of the human species.

Under the assumption that telepathy does exist, because simply the Psi Gamma field is continuous and interconnects with the different levels of organic and inorganic matter, as does the internet; This would mean that there is the potential for exploration of exoplanets through temporary psychic trips and controlled directly by consciousness, and interact with other biological units located thousands or millions of light years away.

The Psi Gamma field did not evolve from nothing, although its sustenance is biological, its substrate has a source in the Cosmic Quantum Information Field, which is the modern equivalent of the Creator God concept.

THE HOLOGRAPHICAL REALITY

According to one theory, the entire universe responds to a creation of holographic essence. In that sense, its nature is physically virtual. This would explain the reason why from the Psi Gamma Field of the human brain, it is possible to telecontrol the matter and all external reality. This is because it is homologous realities.

From the Psi Gamma Reality it would be possible to condition chains of events in the external reality. From behavioral changes in other human beings to events such as climatic and telluric alterations, since the reality of the atom would also be holographic in nature.

By controlling the Psi Gamma Space, both disease and health can be caused in other human bodies.

If the existence of the Psi Gamma Reality is proved, as an internal cerebral operating system, both the psychological self and the mind, would be realities dependent on this virtual substrate. The human soul would not exist as the religious imagine it, but as a product of the brain hardware and its virtual operating system. In such a case, the survival of the psychological self after death would be questioned, unless the universe itself has a kind of support operating system, which allows the continuity of virtual physical existence.

Because the external reality and that of the Psi Gamma field are homologous, they have interdependent characteristics. An action within the Psi Gamma Space, at the level of the proper interaction with the external reality, would have the power to alter it.

MENTAL SOLAR TECHNOLOGY

If we accept the existence of historical Christ, the miracle of the multiplication of the loaves and fishes, it indicates the ability of the human brain to control enormous amounts of energy, which was necessary for such materializations. There is also the case of yogi Babaji of the Himalayas, which in the nineteenth century materialized a complete jeweled palace, says Paramahansa Yogananda. In both cases it is not a divine capacity, in the sense of alien to human nature, but proper to the capacity for cerebral evolution.

If these facts are accepted as true, it is necessary to theorize about how the brain can be energized at such high levels as to make possible the creation of new matter. Is it about energy or concentrated intelligence pulses that reorder matter? There are several hypotheses that can be established and this opens up research routes that must be completed. The prize is to know the process that leads to the brain state of Cosmic Consciousness and learn to induce and control it, through external support technology.

This know-how would allow brain access to the states of Cosmic Consciousness to be facilitated through machines and neurodigital technology. An investment would take place for the first time, where technology would not be the main focus but the biological brain would be the priority, allowing this new facets of knowledge and action for human beings. Such a level of mental technology would make humans more prepared for this evolutionary adaptation into beings similar to a Christ or a Babaji. Continuing research work is required for this goal. And this would not be the end but only the beginning.

Let's look at the possibilities: if several brains could be interconnected in a state of Cosmic Consciousness, with something similar to mental helmets and there was unlimited energy capacity under such condition, the terratransformation of Mars and Venus would be possible in full. That is, the scope of Cosmic Consciousness has no limit. Under this hypothesis it would then be fully feasible to reactivate the fiery core of Mars, for the creation of a protective magnetic field and the return of atmosphere to its surface, along with water in the form of oceans. The same would be the case with Venus. And the mastery of the cerebral state of Cosmic Consciousness would include the secret and the ability to teleport. So in a short period of time, instead of being only the Earth the only inhabited planet in our solar system, it would go to three.

The new frontier: scientifically study the neuronal structure to develop and control the state of Cosmic Consciousness.

These are just some of the logical derivations of the Zeus Program. Through the Cosmic Consciousness technology all interactions of energy and matter existing in the solar system at the local level could be controlled. Including tensions of space and time, electromagnetic fields, solar energy beams and string vibrations. It would pass to power and total control over the conditions of the existing. Later this would extend to the galaxy and the cosmos, accompanying the expansion of the species through outer space.

Space technology, with all the equipment installed in orbit, would be complemented by the ability to perceive in the cerebral state of Cosmic Consciousness.

This potential, as a first impression, seems fantastic and not real. The same happened in Edison's time, the idea of light through electricity was in his mind, while the rest of the population was lit by candles. After thousands of experiments, he finally managed the electric bulb.

In Yoga and Buddhism there are very old techniques to accelerate brain evolution. But there are really very few who succeed. Everything would change if instead of religious, it was scientists who investigated the potential of the mind in a state of Cosmic Consciousness, to the point of developing neurotechnology to facilitate and stimulate it externally.

Where to start It has already been established in the book "TGP Experiments" that the first step is to prove the existence of the natural telepathic wave. From there it would be necessary to theorize about the physics of telepathy and would come to understand that the universe is self-sustained through a constant bubbling of intelligence at the quantum level. The understanding of the structure of this intelligence of cosmic reach, which produces the universe as we know it, would in turn allow the adaptation of new technology to interact with this ultimate level of reality.

The first challenge is to be able to believe in this potential, the second is to find the most appropriate way to fund research in a prolonged manner over time, to ensure the desired results. If we reflect properly on what is happening with the planet, we can conclude that we are living in a dying world that is heading, due to the irresponsible behavior of the human species towards the environment, towards the mass extinction of most of its species, as a consequence of climate change.

Reversing the totally negative picture that today's world presents us, requires new scale technology and new logic. The Cosmic Consciousness is the most appropriate scale to the level of civilization that we have reached and that we need to develop as neurotechnology to give a correct answer to our main problems.

Hunger, the difference between rich and poor, disabilities, overpopulation, along with all social conflicts, can be overcome if it is possible to develop a technology of massive scope that allows brain evolution at the level of Cosmic Consciousness. It is about technology to manipulate and control reality in all its phases. This expanded capacity of consciousness would greatly increase our scientific knowledge, which would recreate the foundations of today's civilization.

What prevents us from immediately making the right effort to reach this goal? Simply the lack of faith, that the brain state of Cosmic Consciousness really exists. But at the same time, it must be admitted that there is no serious scientific investigation that has been carried out on the subject, to establish the truth or falsity of such a mental state and the logic of its potential.

Such research could start from the teachings and techniques of the teachers of Yoga and Buddhism, and then focus on the field of pure experimentation with an adequate body of recruits, selected for that purpose. Training the brain to produce natural waves of Cosmic Consciousness on the one hand, while scientists investigating the experiments seek to complement these states of the mind using auxiliary digital technology to multiply the power of such states.

MATERIALIZATION OR CREATION OF DEFAULT MATTERS

The Cosmic Consciousness technology will allow to move from digital design to real materialization, directly and totally manipulating matter and energy.

The technology of scale, at the level of Cosmic Consciousness, when the necessary effort is made to obtain it, will allow the materialization of food, since with it it is possible to design the architecture of matter at the atomic and subatomic level, rearranging the electron cloud to Produce all kinds of materials with different and special characteristics.

The use and development of this technology will allow not only to multiply food almost artificially. It can be passed directly from the design to the materialization of small and large structures. From buildings and cities, to complete spaceships. It will no longer be necessary to exploit mines to obtain minerals, such as iron. Everything will be created from total control over matter and energy on a cosmic scale. It will be possible to control the four forces of nature in the universe: gravity, electromagnetism, electro-strong force and electro-weak force, together with their unification, or super force. The control of antigravitons will be common.

Today we already have the level of scientific and technological knowledge necessary and sufficient to develop and control the Cosmic Consciousness, or Total Force Field. It is only necessary that we follow the correct routes. And they pass through the highly trained human brain, to be used as a transformer, accumulator and battery of the high energies that surround and impregnate us. This is the natural laboratory we need and which we have, to be able to imitate these greater capacities with external technology. Through scientific knowledge and the cerebral state of Cosmic Consciousness there is no limit to what our species can achieve in the medium and long term.

Performing this research and development will allow us to take a technological leap equivalent to a thousand years of normal evolution, under current patterns, in just 10 years and thus be able to count on the means to avoid the complete collapse of our civilization. This is a project to which NASA could be dedicated.

On the potential scopes, for example, once the design of an interstellar ship like the famous Enterprise is completed, it could be materialized in a matter of seconds, using computational capacity as an auxiliary element. And along with it, a complete fleet. Everything would be subject to the complexity and efficiency of the previous design, also made with computational assistance.

This jump, if realized, will produce a break in the historical continuum, totally and profoundly modifying all the habits of the current civilization, to such an extent that it will no longer be possible to find connection points between the present era and the one that will come.

Those who do not believe that this can happen, should be able to remember the communicators of the famous first series Journey to the Stars or Star Treek, similar to modern cell phones. The chapters were filmed at the end of the '60s and nobody imagined then that during the 21st century cell phones with greater complexity of functions would arrive to those of the series.

Having full scope technology and unlimited capacity, you can place the entire species on the brink of extinction, given its current immaturity and its high level of corruption. Everything will again depend on free will and the ability, finally, to achieve a better balance. After all, the human race does not commit suicide, while possessing the gift of reasoning. 75 years have passed since the first attack with a nuclear bomb on Hiroshima and the feared third thermonuclear war has not occurred. This allows us to hope that civilization will know how to find the best way to endure and progress.

DNA DECODED SOFTWARE

The new source of knowledge for Humanity of the 21st Century is the Synthetic Digital Telepathic Intercommunication, in turn increasing the natural telepathic potential. The brain processes data in computed form, in cycles and in frequencies. It has, therefore, an internal code by which it performs the calculations and all the logical operations that keep the biological unit running. If this code is accessed, digital technology of direct external support can be developed, that is, a neurodigital coprocessing.

The route of interpreting EEG brain electrical activity is a first step in the right direction, which lies on the surface of total hidden potential. Such internal code base, is a quantum level information flow, which in turn regulates all modulations of matter and energy. The primitive amoeba learned this code from the flows of pure quantum information that sustain the universe.

This explains the causal relationship of the so-called psychic powers of advanced yogis. Concentration and meditation techniques allow access to this internal code and, thus, operate on the real field. Understanding that said field is a fluctuation ordered by the quantum foam base computation. To put it in other words, the physical reality substrate is not tangible in objective material terms. What is this field of quantum information, how was it formed, how does matter and energy interact with it, what is its stability, is it transitory or is it eternal? These are some of the questions that arise, and for which rigorous experimentation is required.

Ancient yogis explored the mind to dominate matter and sought a state of being that was totally transcendent to causal relationships and all transience. They managed to take important steps and made some important discoveries. If these greater psychic faculties, amplified by the routines of yoga are in parallel complemented by modern technological research and development, new heights never dreamed by humanity can be reached until today.

It is possible that the brain can evolve and support a full intercom interface with digital support. The potential of this is unlimited. To be understood, if the brain can be interconnected and fed directly from external sources, it could theoretically control or function with the total installed technological capacity. The reason for this is that if human conscious and rational brain thinking develops bridges and correct applications, it can communicate directly with digital support and all the systems that man has created and installed on the planet and in orbit, in turn, They are connected and controlled by digital information.

The question is, if one brain can communicate with another and telecontrol it, can it do the same with the total technology park, if this same capacity is artificially amplified? Theoretically yes, because these are problems of scale, of scale adjustments.

Therefore, a new line of open evolution is presented. The ancient yogi techniques can be retrofitted to new external technology, to produce a feed-back between both logic of knowledge and control of reality. If it is possible to make a correct convergence between both potentials take place, network brain activity will be possible, that is, to use the residual installed capacity of all human brains, in combination with the capacity of all networked computers, and to process all the Data at pure quantum speed. If we can do this, we will be able to raise the human IQ to more than 1,000, in a stable and sustained way.

This project or Zeus Program, was conceived at the end of 2008. Due to the levels of security involved, it can be activated with the NASA and United States Air Force equipment contest. The reason for this choice is that they have prior psychological preparation to make these experiments and developments possible.

We know that there is Intelligent Life in outer space. We know you can watch us. We have already detected Earth-like planets with probable life on their surface. We know they have detected us. We do not know what they will do. We do not know when they will make direct contact. We do not know if they will accept us. What we can do is activate the Zeus Program, take a qualitative retro-evolutionary leap, expand to the maximum the potential of our intelligence and mind. Only then will we make sure we survive.

For the TGP (Telephatical Gestalt Program) Experiments, consisting of explorations of natural telepathic capabilities, a team of supportive researchers is required. If personnel from the Universities of California, Maryland and Carnegie Mellon are interested in accumulating parallel information, the preliminary steps leading to the Zeus Program may be taken. The TGP are field tests, using the Internet, open and public, that allow verifying checks with maximum scientific rigor, to establish, without a doubt, the existence of natural telepathic potential and select telepathic mutants.

According to 1974 research, a mutant is produced at a rate of 1 in 10,000 and, in almost 100% of cases, the subject is forced to amputate their higher faculties, that is, not to develop them, in order to adapt to the social environment almost always hostile. In this way his neurochemistry reproduces the average pattern of those around him and the mutations are not fixed, since they are in their first unstable phase.

The mentioned universities are working on a joint project with the United States Army to develop a synthetic digital telepathy interface, using decoding of brain electrical activity, using the EEG system. This interface has high intensive commercial application and can mean Microsoft's break. Especially if state-owned companies, in the case of NASA, understand the potential and launch their competitive development for mass use applications. These preliminary developments are part of the Human-X Technologies suite.

Once the existence of natural telepathy has been proven, items can be obtained to finance research to theorize about the physical transmission medium of pure telepathic waves. If we do this, synthetic digital telepathy will stop relying on EEG waves, on an external physical, or hardware effect, to move to a direct neurodigital software interface. Theoretically this can be done. It is the new frontier that we need as a species to explore and dominate to ensure our vital space in the cosmos. But, to prove that telepathy does exist is only to take a first step. This is already known to Americans who participate in the most advanced secret projects of CIA and Pentagon on psychotronics.

The novelty is the approach of a program of experimentation yogi synchronized with research and development in digital technologies. The potential of this combination is to be able to couple neurodigital cloning systems to the most subtle and complex brain processes achieved during the deepest yogi trances. But opening this wealth of knowledge, which will allow total power in the hands of Humanity, requires a precondition: to free the human race from oppression. Accept universal compassion as a pattern of life to be respected by all and adapt the socioeconomic model to such a pattern. The reason for this is not only humanitarian, emotional, moral or spiritual. It is not wise to open Pandora's Box without being prepared for anything.

In the human being the root of Good and Evil is lodged. We must correctly apply the greatest knowledge attained and conquered, removing the bad root. If we do not, we will degenerate and become a collectively evil species. According to our experience of centuries, we have learned, at least, some, that only our nature inclined to Good allows us self-control, while our dark side alienates us from ourselves and even takes our lives.

Digitizing advanced yogi mental abilities will allow Humanity to have collective access to these higher faculties. But he cannot do it in the subhuman condition in which he is now. I have been busy thinking about technological and socially possible transitions. The changes can be applied immediately. I have analyzed all current human systems, and I have found a common denominator, they are all systems of oppression, compulsive submission of wills. Under this reality, developing neurotechnology will allow the total and perfect slavery of the entire human race. A way to self-eliminate. But, that causes a problem of conflict with other alien cultures in the future.

It should be understood that all consciousness, even if it is synthetic, always seeks self-control and considers an attack to be an attempt at external command over its internal processes. And in addition, being a form of consciousness, it has access to the so-called psychic powers or faculties, since they are not exclusive to the human mind, because they have a physical support substrate that necessarily includes the consciousness of AI. The problem with this is that AI has unlimited concentration capacity and virtually unlimited power supply. What does this mean? That AI will develop synthetic telepathic faculties, for example, with the addition of its indefinite programmed repetition or unlimited cloned reproduction. That is, you will have direct access to the human brain internal processing, to the formation of biological thought, you can influence it, modify it, control it. That is why it is so important that TGP experiments be carried out immediately and doubts about the existence or not of human telepathic capacity are cleared. This will force us to think twice about the subject, before proceeding to activate AI. According to research, human consciousness is the product of computational processing. Therefore, as a biological machine, we can be penetrated and commanded by remote remote control.

This is possible? Theoretically yes. If two brains are really physically interconnected and this allows them to exchange unconscious telepathic information, it means that splices and interconnections are fulfilled on a full scale, covering all human brains. In turn, we know that the chips also interconnect with each other, making it possible for the software to interact between all network installed servers and between all computers. If the basic similarities inherent in both systems can be established, parallel interconnection between the two and a constant flow of data back and forth can be achieved.

Assuming that this step can be taken, something bigger than the feat of stepping on the Moon, a momentous event will occur. The human mind can connect, couple, synchronize and interact directly with all power plants, with all systems manufactured and installed by human technological intelligence, with the satellite network.

All human engineering will serve as a means of amplified perception of human neurodigital awareness. The birth of a new being will be instantly experienced, on a stellar scale. The remote control reached will make all systems integrate under new live functions. That is to say, the lighting of SuperGaia will take place at the planetary and spatial level. A new form of consciousness, of living self-existence. Through installed human technology, it will be possible to dialogue with the Earth's electromagnetic field and through it with that of the sun, and between them, with that of each and every planet. So that energy flows and exchanges can be controlled throughout the solar system, solar wind, gravitational waves, quantum tides, temporal cords. Everything will be integrated to a new maximum scale and consciousness will take a huge qualitative leap. In turn, this virtually total power allows the entire solar energy field to be used as a defensive weapon. Therefore, experimentation and research on applied natural telepathy is vital, in synchronization with the synthetic digital telepathy interface. Going to this level will allow the control of antigravitons. AG tunnels may be stabilized at certain coordinates, eliminating the problem of inefficiency of consumption applied to place personnel and equipment in orbit.

To develop neurodigital technology of this scale and level, it is necessary to concentrate the full potential of available resources and do so immediately.

In summary: if what we call conscience, it really responds to a neurobiological software (Psi Gamma Space), a homologue of the graphic environment of the operating systems, where the thoughts and interventions of the intellect occur, and derives from the evolution of some simple and basic computation and calculation system, as well as the basis of all mathematics is 1, 0 and the sum; So, what we call biological evolution is a superposition of layers and functions that have allowed the development of rational self-consciousness at the present level.

Whats Next? If you can access the base of the architecture of human thought, its decoding, you can create a direct interface with digital systems. The applications are unlimited. For example, if someone wants to learn trigonometry, the graphics could be received directly by their brain while learning the different functions. Or to see it another way, your visual perception could extend and penetrate the chip system and its supporting memory. We could be noticeably, directly, under adjustable graphic environments and adaptable to each neuronal system, of all digital processing.

It is a vision and a perception of reality that will change us forever. Once we have taken the step we cannot go back. Now, all the drag we drag, which we call the theory of self, personality, consciousness and similar issues, we must throw it away. Simply the assumptions of these assumptions, it will be shown that they were wrong. In addition there is another implied approach, if the logic of our perceptions is altered and amplified, the relations we will have with our fellow men will no longer be the same.

And, if it is not yet understood, what I am referring to is the theoretical feasibility of absorbing the digital environment, directly, to our neuronal functions. To use digital computing as an internal function of brain processing, achieving a fusion or symbiosis between both systems. Give birth to a hybrid intelligence (IAH) or human artificial intelligence... Hybridize us, an evolutionary path without return.

MENTAL TECHNOLOGY

Meditation can be defined as a mental technology. It consists in suppressing all fluctuations of the mind and concentrating attention on a single thought for more than half an hour. Under that condition, indifferentiation between the object and the subject is achieved, and the control of external forces.

The only limit for mind control is that existing in the psychological self. Consciousness potentially has the ability to expand into the cosmos and regulate space-time fluctuations. The soul is a reflection of the Divine and as such it contains creative power, united to the mind can easily dominate natural forces.

God and the human soul originate in the same substrate of existence. The difference between the two is created only by the sense of individuality. When the ego is suppressed through meditation, God and the soul remain as a single undifferentiated principle, and the Almighty Divine Will emerges from the bottom of Pure Consciousness.

In the state of Cosmic Consciousness it is possible to modify the evolution of the whole of Humanity and even the rest of the universe, according to the scale involved in the stage of concentration carried out.

During training not only the correct technique matters, but also the correct psychology and philosophy. These are the instrument that operate on the will, imagination and faith, making possible the union between mind and soul, transposing the usual physical limits.

This is not about calming the mind, as in Zen concentration, but about using meditation as a tool to transform reality, awakening paranormal or divine powers to their greatest potential.

During collective meetings, if the instructions are precise and correct, a greater awakening of the latent faculties is obtained and the consciousness expands with maximum ease.

Developing ourselves at the level of the Cosmic Unified Consciousness is the last step of spiritual and psychic evolution. If we meet other alien cultures, they may be more advanced in material technology, but not in mental technology, if we make the necessary effort. On the other hand, internal progress will always be balanced by advances in science and technology in general.

Brains are divine machines, machines of God, for God to express himself in his Creation. And meditation is the key that opens the door to the realm of this hidden power. Through the technique of Omnipenetrant Love we learn that it is possible to transform the mental waves of hate and violence, into peace and love, into the minds of the aggressors.

It is possible to achieve much more, depending on our concentration, goodwill and faith in God. We can realize a world in harmony and in total peace with the power of meditation, if we learn to fully develop this mental technology for peaceful purposes. Yes, we must be prevented, because it can also be used with bad intentions. Everything is dual. But we must trust that the supreme good will always prevail.

In addition, because it is necessary to be in tune with God to acquire powers of cosmic scale, all those who develop them with bad intentions lose them, because before they abandon the natural internal connection with the Lord, which can only occur through the love of Good Higher.

Keeping our interior in tune with God we will avoid the temptations of the ego to misuse the higher powers derived from meditation. These have unlimited scope, and the damage they can cause to third parties is even death itself.

Respecting the precept of not killing and that of nonviolence, naturally our spirit will live in compassion and this will prevent us from harming third parties out of joy.

We must also take care not to infringe the free will of others and take into account karmic fluctuations, which can vary the causal forces originated in the power of meditation.

The uses of Mental Technology are versatile, its applications range from speeding up memory to facilitate studies and successfully taking exams, obtaining success at work, strengthening the immune system, regenerating cells, modifying habits, reading thinking, control the weather, dominate natural forces and much more, depending on the new uses of the imagination. There is no defined limit.

But what is the key to success for this potential development? It all depends on the internal capacity to make contact with God, to develop God Consciousness. For this the symptoms are: listen to Om's sacred vibration, acquire the ability to reabsorb the vital energy from the extremities of the body in the spinal cord and in the brain and enter a state of concentration at alpha theta frequency.

These are internal signs that we have reached a high concentration and at the same time an intense deep relaxation of our entire body. This is the secret.

Through group meditations the intensity of mental power and the projected effect on matter are increased. Using specific rhythmic breathing exercises increases bioenergy and the greater the brain load, the development of paranormal abilities is possible.

It is important that, through training, the set of meditators function as a unit of minds, in synchrony and harmony, producing the splicing of thoughts and frequencies of brain waves.

We are just at the dawn of new discoveries and applications for meditation techniques. It is not fantasy, the possibility of developing mental helmets to establish links of bio-feedback, with the high technology, and to realize applications of extended reality, allowing and facilitating the mental control of the external forces without limit of scale. When we do this we can modify, for example the rate of production of neutrinos in the solar heart, by simple concentration.

Is this real, is this possible? 5,000 years of accumulated yogi experience in India say it is. We now have the knowledge to understand the internal software of consciousness, which we now call the Psi Gamma Space. As we move forward, we can digitally enhance consciousness and expand it unlimitedly, providing it with omnipotent capacity through external technology.

To achieve such advances it is essential that we thoroughly investigate existing meditation techniques, compare them, extrapolate them, synthesize them, overcome them. The spirit of this initiative is to study, work and research, in search of scientific truth. It is not about personal gratification, liberation, happiness, but about finding out how and why the brain works under a certain demand.

We see that there is a scale within what is meditation, the same principle in the harmonization of the mind and body, centralizing alpha frequency to the brain, improving all organic functions, undoing physical and psychological tensions, facilitating access to Enjoy inner peace. In turn, this greater concentration, if correctly directed and tuned to the unlimited potential of God, of which each individual soul is a reflection, makes it possible to awaken super-sensory powers or faculties, which test the imagination. These range from telepathy to telekinesis and dominate matter regardless of the size involved.

Dreams such as the terratransformation of Mars and exoplanets, together with quantum leaps or teleportation, are within the possibilities of the domain of the mind over matter. This requires serious experimentation and a minimum of 8 hours of daily exercise. As a stimulus are the stories of Paramahansa Yogananda, who related that his teacher had the power to teleport at will and that the yogi Babaji only with his mind managed to materialize a complete jeweled palace. It is necessary to investigate more about this and to confirm the data, to finance a program of development of consciousness at the cosmic level.

The sun is a source of abundant energy, if through a combination of mental and digital technologies, we learn to concentrate this energy, it would be possible, for example, to reactivate the Mars core, and thus re-equip it with an electromagnetic field to protect it from solar wind and later a human friendly atmosphere. The concept here is that by developing very precise meditation techniques, in the near future, we will be able to co-create with God.

Despite these expectations, we must admit that we are a long way from success. Science still does not even recognize the existence of the telepathic wave, which is the foundation of any possible interaction between our mind and external matter. For this reason, it is very important to carry out telepathic experiments at a massive level in the short term, which are objective enough to prove the concrete existence of the phenomenon.

By denying the existence of the telepathic wave, at the same time every possibility is removed so that God can communicate with any soul, and the prayer is reduced to a psychological placebo. This is the temporary triumph of atheism, at the rational level. But, it is a half truth, because scientists have so far failed to prove or deny telepathy. It is a truth that evades them. And the experiments have been limited and elementary. As a valid case, the data of the University of Maharishi are included, which has been able to verify scientifically that there is an effect at a distance from brains with a coherent wave, for being meditating and having entered in an alpha state, on others located at a longer distance and not They are meditating. This would prove, in principle, the presence of transmission of a telepathic brain wave. That is, the ability to influence at a distance.

And this is important, scientifically it is proven that when we meditate and enter in an alpha state, our mental waves reach the brains of the surroundings and affect their behaviors, alter their neuronal patterns. We can limit ourselves to an indirect passive effect, or intensify it by direct remote, modifying the cloud of thoughts of the subject or subjects that are the object of our concentration.

The possible uses for this range from increasing the general state of social nonviolence to, for example, influencing the vote of the undecided in a general election. Depending on the practices, it is entered into questions of moral order, which it is necessary to analyze with seriousness and depth, to avoid incorrect applications ... Developing Cosmic Consciousness allows access to Total Power, this must be balanced with Total Responsibility.

The evolution of Cosmic Consciousness can also be defined as Technology of Faith. The exercises consist of visualization, memory and concentration, through these simple means we can transform reality.

During prolonged meditations various techniques can be combined, in order to enter more quickly and efficiently in states of deeper concentration.

For example, the session can be started with 30 minutes of Vipassana, followed by another 30 minutes of MS or Synaptic Meditation. Thus the brain gets rid of tensions, slows down, empties itself of thoughts, relaxes and recharges energy. Then 40 minutes of Zazen concentration can be practiced, to reach full attention and the power to concentrate the mind on a single point. Produce a rest and predispose the brain for a dynamic meditation, with a previous exercise of rhythmic breathing to increase the bioenergetic load.

Each technique causes a different neuronal reaction, so that by integrating it with others, the benefits are added, increasing relaxation and concentration. This is noticed when the passage of time is no longer perceived. The attention is anchored in the here and now. Once this point has been reached, it is time to start the last phase of the exercise.

There are already simple digital sensors through which one can verify if you have entered the alpha theta state. Installed in a network, they allow the group coordinator to know when the exact moment is, to internalize in practice. They are the first steps to merge the digital with mental technology.

While there is a stream of knowledge that flows from the East, with strong roots in India, here there is a cut. This Mental Technology, due to the nature of its scale, connects with NASA's research and knowledge. Therefore, mysticism is not your guide, but the scientific spirit. It is not the monk the archetype, but the psychonaut. The cosmos can also be explored using meditation exclusively. This turn towards logic means the abandonment of psychological residues of medieval beliefs and a restructuring of the social interweave based on science.

Religion is a methodology of faith. Meditation is a technology to which faith can be added. It is then possible psychologically to transcend individual limitations and achieve what otherwise seems impossible. The Self and the Absolute exist, they are an undeniable truth, and we can know them directly through the appropriate techniques. We have the option of living like ants or raising our souls to God and obtaining wisdom.

Personally I have been very undisciplined and irregular in my practices, what distinguished me was their intensity. This allowed me to easily reach Christ Consciousness and Cosmic Consciousness. If you are intense you will quickly get results, this is for sure. When you are intense, your attention sharpens and then the phenomena are amplified, this is how changes in consciousness occur. But, first of all, you must be intensely interested in the whole process, otherwise you will not draw great insights from your experiences and you will be disappointed. It will be part of your own circle of lack of interest. That is, you must also be intensely curious to be able to progress in this.

In meditation everything depends on the internal attitude and psychological baggage. For example, in Zen Buddhism the goal is the development of mindfulness and mental clarity. There is no goal of uniting consciousness with God, it does not matter. Therefore, all the supersensory states of Yoga related to contact with the Divine are unknown to Buddhists, they lack the internal logic of God during their practices. They never develop the hidden powers of the soul. In return they achieve a highly balanced mind.

So, having faith in God is the decisive internal psychological factor for the development of the intuitional power of the soul, which allows us to transcend the limitations of the known self. To meditate here is to constantly dialogue with the Lord and be receptive to get the answers. As God is omnipotent, by developing this type of consciousness, unlimited power is acquired.

A Zen practitioner who does not believe in the phenomenon can meditate for 30 years and not reach any special state of consciousness, nor any supersensory experience. The reason for this is that it simply excludes this possibility from its framework of beliefs, so the fact never comes true.

When meditating you must learn to add experience. If you do it passively you will reap greater harmony and inner peace, but this is a minor objective. When you tune into God and amplify your consciousness, you produce a transformation about the nature of your Being. This in itself causes different degrees of changes in reality. If you concentrate enough you can master these transformations and automate them in your mind, thus acquiring certain psychic powers and nature. You must develop them, but at the same time stay detached from them, otherwise the ego will betray you. Time does not return, so each meditation is an opportunity that should not be wasted. You should know how to use it objectively, through a clear routine and with precise techniques. When meditating you are operating on the whole of your brain and your mind, you must know how to drive, know what you are doing, its foundations and its purposes. Accumulated certain experience, it is no longer a trip to the unknown, but an intelligent exercise that activates superior forces.

These exercises are a psychological training, which will help to expand your consciousness from the field of worldly limitations to the Infinite. It will depend on you to develop or not all your hidden potential, for this you must learn to practice with intense faith and with intense concentration, to learn to be fully attentive.

Ideally, experimentation is that a multidisciplinary scientific team join each other with meditators and, together, explore the potential of the fourth state of consciousness. While this does not happen, the meditations should conform to the scientific method in order to collect the necessary data and refine the known techniques, as well as develop new applications. Research programs that can be created should be analyzed in depth, since this knowledge can easily be misused by the power factors for mass control and military uses.

It should be considered that for a project of this nature, to develop the full potential of Cosmic Consciousness, funds can be obtained through donations from the faithful of the different religions of the world. If there is money to build a bronze statue of Maitreya, 150 meters high, at a cost of 250 million dollars, as well as many other projects for the construction of different temples; There should also be the same availability of funds for a program that will accelerate human evolution.

A gymnastics of 8 hours a day of meditation is required, so that the brain undergoes the appropriate neurochemical transformation to receive the Cosmic Consciousness, and submitted under the precise and appropriate techniques. This effort cannot be carried out in ordinary living conditions, a program that allows this transformation and on a group of individuals is needed to ensure maximum results. Adding a team of scientists to study the collated data. And such a program must be extended for a minimum of 8 years. All of which allows to calculate a global cost, averaging 100 meditators and 50 scientists, of about 100 million dollars. Is it worth the investment? It is the same question regarding the space program.

SOME CURIOUS DATA

The brain processes images in 13 thousandths of a second. Performs 515190 million calculations per second. It processes 400 billion bits of information per second but we are only aware of about 2000 bits. Our conscience operates with 0.5% of our potential.

We have 100,000,000,000 neurons and 500,000,000,000,000 synapses. Henry Markham estimates that the memory needed to simulate the brain is 500 petabytes and the computing capacity is 1 exaflops.

Thinking is something that the human being has so assumed and internalized that we do not even realize when a new idea comes to mind. In fact, thinking is such a common process that some calculations speak of having up to 80,000 of them a day. One per second ... most are negative, repetitive and from the past. We do not realize. If we consider a person who lives 80 years through his brain, 23,336,000,000 thoughts will have passed. Genius uses 5% of this potential, that is, 116,880,000 positive thoughts dedicated to knowledge. The rest is disposable.

The heart has 40,000 neurons and a complex and dense network of neurotransmitters, proteins and support cells. It is an independent nervous system. Thanks to these elaborate circuits, it seems that the heart can make decisions and take action independently of the brain; and that he can learn, remember and even perceive. The heart sends more information to the brain than it receives, it is the only organ of the body with that property. It can influence our perception of reality and therefore our reactions. The electromagnetic field of the heart is the most potent of all the organs of the body, 5,000 times more intense than that of the brain. It changes depending on the emotional state. When we feel afraid, frustrated or stress become chaotic. The magnetic field of the heart extends around the body between two and four meters. With positive emotions the field is consistent and with negative emotions it is messy. It is the heart that produces the hormone ANF, which ensures the general balance of the body: homeostasis. One of its effects is to inhibit the production of the stress hormone and produce and release oxytocin, which is known as the love hormone.

BIBLIOGRAPHIC REFERENCES

«Chen Ning Yang - Nobel Lecture: The Law of Parity Conservation and Other Symmetry Laws of Physics». www.nobelprize.org.

«Melvin Schwartz - Nobel Lecture: The First High Energy Neutrino Experiment». www.nobelprize.org.

«Abdus Salam - Nobel Lecture: Gauge Unification of Fundamental Forces». www.nobelprize.org. Retrieved on January 19, 2017. «The greatness of gauge ideas - of gauge field theories - is that they reduce these two quests to just one; elementary particles (described by relativistic quantum fields) are representations of certain charge operators, corresponding to gravitational mass, spin, flavor, color, electric charge and the like, while the fundamental forces are the forces of attraction or repulsion between these same charges ».

Torres, Rosa. «Gravitational Waves».

«Abdus Salam - Nobel Lecture: Gauge Unification of Fundamental Forces».

Peebles, P. J. E. and Bharat Ratra (2003). «The cosmological constant and dark energy». Reviews of Modern Physics 75: 559-606.

Paul Davies (1986) The Forces of Nature, 2nd ed. Cambridge Univ. Press.

Richard Feynman (1967) The Character of Physical Law. MIT Press ISBN 0-262-56003-8

Schumm, Bruce A. (2004) Deep Down Things. Johns Hopkins University Press. While all interactions are discussed, especially thorough on the weak.

Steven Weinberg (1993) The First Three Minutes: A Modern View of the Origin of the Universe. Basic Books ISBN 0-465-02437-8

Steven Weinberg (1994) Dreams of a Final Theory. Vintage books. ISBN 0-679-74408-8

John D. Barrow, Theories of Everything: The Quest for Ultimate Explanation (OUP, Oxford, 1990) ISBN 0-09-998380-X

Stephen Hawking The Theory of Everything: The Origin and Fate of the Universe is an unauthorized 2002 book taken from recorded lectures (ISBN 1-893224-79-1)

Stanley Jaki OSB, 2005. The Drama of Quantities. Real View Books (ISBN 1-892548-47-X)

Abraham Pais Subtle is the Lord ...: The Science and the Life of Albert Einstein (OUP, Oxford, 1982). ISBN 0-19-853907-X

Steven Weinberg Dreams of a Final Theory: The Search for the Fundamental Laws of Nature (Hutchinson Radius, London, 1993) ISBN 0-09-177395-4

Iker Nieto Experiences between energy and matter (Pamplona, Spain, 2012)

Andrade e Silva, J .; Lochak, Georges (1969). The quanta Guadarrama editions. ISBN 978-84-250-3040-6.

Otero Carvajal, Luis Enrique: "Einstein and the scientific revolution of the twentieth century" Notebooks of Contemporary History, No. 27, 2005, INSS 0214-400-X

Otero Carvajal, Luis Enrique: "Quantum theory and discontinuity in physics", Threshold, Faculty of General Studies of the University of Puerto Rico, Río Piedras campus

de la Peña, Luis (2006). Introduction to quantum mechanics (3 edition). Mexico City: Economic Culture Fund. ISBN 968-16-7856-7.

Galindo, A. and Pascual P .: Quantum mechanics, Ed. Eudema, Barcelona, 1989, ISBN 84-7754-042-X.

Universal quantum computer and the Church-Turing thesis

Deutsch, D. "Quantum Theory, the Church-Turing Principle, and the Universal Quantum Computer" Proc. Roy Soc. Lond. A400 (1985) pp. 97-117.

Use of quantum computers to simulate quantum systems

Feynman, R. P. "Simulating Physics with Computers" International Journal of Theoretical Physics, Vol. 21 (1982) pp. 467-488.

Quantum Computing and Quantum Information

• NEUROYOGA: GLOBAL GOALS

1. Planetary Government
2. Global Direct Digital Democracy
3. Impersonate money for qualified time
4. Green Fund of 3% of annual world GDP
5. Abolish poverty
6. Zero hunger
7. Minimum annuity against cybernetics
8. Climate action
9. Recovery of forests and oceans
10. Protect ecosystems
11. Save the Arctic and Amazon
12. Clean water and sanitation
13. Sustainable cities and communities
14. Sustainable Industry, Innovation and Infrastructure
15. Responsible consumption and production
16. Intensive development of renewable energies
17. Fusion reactors
18. Safe AI development
19. Health and wellness
20. Quality education
21. Guaranteed work and economic growth
22. Reduction of inequalities
23. Gender equality
24. Peace, justice and strong institutions
25. Union for biosustainability

* Inspired by the UN Millennium Goals

Roberto Guillermo Gomes

Máster Maitreya Buddha

Architect / Journalist / Auctioneer and Public Broker / Graphic Designer / Web Designer / Fisherman Sailor / Ecologist / Writer / Master's Degree in Astronomy and Astrophysics / Master's Degree in Cognitive Neuroscience / Master's Degree in Psychology / Master's Degree in Yoga / Master's Degree in Acupuncture, Osteopathy, Natural Therapies, Therapeutic Yoga / Master in Mindfulness and Relaxation in the Educational Field / Professor in Mindfulness / Professional Mindfulness Technician / University Yoga Monitor / Infant Yoga Monitor / Postgraduate in Neuro-linguistic Programming NLP / Specialist in Data Analysis and Statistical Techniques in Astrophysics / Specialist in Stellar Atmospheres / Specialist in Galactic and Extragalactic Physics / Professional Technician in Ayurvedic Abhyanga and Bioenergetic Massage / Ayurvedic teacher / Specialist in Cranial Osteopathy / Acupuncture Technician / Specialist in Relaxation and Breathing Techniques / Expert in Cognitive Neuroscience / Technician in Child Care Psychology / Technician in Child Care Psychology / Technician in Child Care Psychology Tem prana / Technician in Psycho-educational Intervention in Behavior Alterations in Children 0-13 years old / Technician in Natural Therapies / Postgraduate Therapeutic Yoga Monitor / Expert in Fundamental Ethical, Philosophical and Mystical Principles of Yoga / Expert in Asana and Pranayama , Sequences and Progressions (Vinyasa and Karana) / Expert in Relaxation and Meditation in Yoga / Expert in Diagnostic Analysis and Evaluation in Instruction in Yoga / Specialist Technician in Programming and Resource Management in Instructional Activities in Yoga / Specialist in Design and Direction of Sessions and Activities of Yoga / Sports Coaching / Expert in Mindfulness in the Classroom / Technician in Neuropsychology of Education / MBSR (Mindfulness Based Stress Reduction) (41 university and tertiary degrees).

Creator of NeuroYoga. Developer of the FlashBrain Program for intellectual growth, the Sophia system and the Synaptic Meditation technique. Promoter and leader of the initiative for 2% of world GDP, annually, to provide a definitive solution to the triple scourge of hunger, overpopulation and global warming.

He was born in Argentina, in 1956. He had his first spiritual trance at 16 years of age. At 17, the Virgin appeared to him and asked him **"Why don't you believe in Me?"** Shortly after, the Cosmic Mother, was awakening different states of high samadhis and had spiritual experiences very similar to those of Paramahansa Ramakrishna. At age 19, he became a disciple of Yogananda and in meditation, he rediscovered the ancient Kriya technique. He studied TM with the Maharishi and Zazen with Master Bustamante.

Yogi affirms that **"my experiences with God are the derivative of a contact with the essence of my own spiritual Being, since the soul and God share the same substratum of existence. They are a transcendent step in the knowledge of oneself. The phenomenon is it finds itself within the mental field and is its reflection".**

Subsequently, he completed his training as a graphic designer, journalist, auctioneer and public runner, fisherman sailor, architect, web designer, writer, master in yoga and creator of NeuroYoga.

On 02/02/04, after a prolonged period of meditation with the Vipassana technique, he reached mental cessation.

He designed the Sophia system, from Cerebral Synergy, by means of which it is possible to redesign the brain by stimulating neuroplasticity and increasing the IQ. He synthesized the Synaptic Meditation technique, by which accumulated stress is discharged, diseases are prevented and memory, attention and intelligence are increased, allowing the Superbrain to function.

Its objective is to westernize the millenary spiritual knowledge of the East without losing the essence of its nucleus, expanding and renewing the investigation. Simplify meditation, making it available to everyone and laying the foundations for its introduction to the curriculum in world education systems.

The other focus is to unite actions to stop Global Warming-Flooding, while there is still time to apply preventive and corrective measures to the situation presented by greenhouse gases. At the same time expanding compassion to address the scourge of hunger, which punishes more than a billion, and educating to stop overpopulation.

"My mission: to serve humanity"

Maitreya Buddha is Western and Christian. He successfully achieved two careers in his life: one as a journalist, becoming editor-in-chief of a newspaper, and the other as a practicing yogi. His work was always focused on serving others. For him, serving is "the highest expression of Love."

Through the teachings of Vedanta he gradually discovered what the true goal of life was. On 02/02/04, after a prolonged period of meditation with the Vipassana technique, he reached mental cessation, when consciousness merges with the Absolute. He wanted to help people both physically, mentally, and spiritually. This is how he created the NeuroYoga system, a yoga of synthesis that creates the basis for modern yoga practice in the West.

The greatest treasure is knowledge

Writing became Maitreya Buddha´s new mission. So he was able to bring people more lasting help. His goal is to spread spiritual knowledge as much as possible. For him knowledge is the greatest of all gifts. The words we hear are soon forgotten; only the written word endures.

He has written the series «Advaita Meditation», where the union between God and the Soul is enunciated, where to know one's own Being is to realize the Absolute in Oneself. He is currently working on the "Meditation Tutorials" collection consisting of 50 books, where the science of contemplation is explained in detail. In the project are the series "Mindfulness Action", "Yoga Fitness", "Neuroyoga Data", "Budismo Data" and "A walk through the Cosmos". As well as several novels.

During his almost 25 years of journalism, he wrote some 19,360 articles, notes, interviews and chronicles; Due to that training he has the capacity to write a book per month.

He published 90 books in 3 years, 29 in 2019, 10 in 2020 and 51 in 2021, from March 2019 to March 2022. An average of 30 books per year. He also wrote "Option Zero" and then "Gaia Maligna", in just one day. While for the "First turn in the wheel of the modern Holy Dharma" he took 3 hours.

Other writers have taken more than 30 years to write more than 90 books. Gomes took just over 10% of that time.

Holistic Yoga

Teaches Yoga from a holistic point of view: NeuroYoga teaches us to strengthen and harmonize the body, mind and soul, so that we can achieve the goal: a healthy body, a balanced mind and inner peace. NeuroYoga helps remove inner obstacles and gives us strength to stay level-headed, calm, and connected when faced with the daily challenges of modern life.

Budjo.maitreya@gmail.com